Title: RARE METALS - The secret ingredients of our future

Author: Roberto Gozzetti

Copyrighy © 2021 FREE2READ

Translator: B.B.

PREFACE

This book was born thanks to the precious contribution of all those readers who over the years have followed my journalistic activity on metals. Their questions, their curiosities and their comments in learning about rare metals led me to collect and rearrange a series of articles published over time, to offer an organic writing that could satisfy the curiosity of other people.

The author

INTRODUCTION

Talking about rare metals to a wider audience than metallurgical engineers may seem a little crazy.

However, during the many years (I will not tell you how many so as not to allow you to calculate my venerable age) in which I have dealt with rare metals, I have always found a great interest from people who had nothing to do with metallurgy or with the metal sector in general. Perhaps, the fact that it was something rare and unknown, a bit like when we talk about mysterious animal creatures, sparked an instinctive curiosity in people.

A curiosity that grew even more when talking about the extraordinary properties of some of these metals and their use, for example, in the military or technological sectors.

On the other hand, how not to be fascinated and frightened by a metal like polonium, radioactive and deadly, which jumped to the headlines because it was used to liquidate an inconvenient spy in the Kremlin. Or from Californium, recently discovered (1950s) and potentially usable to build pocket atomic bombs. But I would not like to anticipate you too much, taking away part of the taste for reading.

Two more words just to underline that the book is not a technical manual of metallurgy, nor a guide to the trade of metals, rare or less rare. Instead, it is a journey into the world of rare metals for all those readers who want to understand what they are, how they are used, the commercial background and the big players in the market, what importance they have in our society now and in the coming years, these extraordinary "secret ingredients of everything", as the famous National Geographic magazine correctly defined them.

UNKNOWN, TERRIBLE AND PRECIOUS

A METAL FOR POCKET ATOMIC BOMBS

Ten billionths of a second are enough to put an end to over two centuries of history, less than a moment.

A frightening explosion whose heat instantly causes the volatilization of every organic substance, making every living thing disappear over a period of kilometers. And these are only the thermal effects of an atomic bomb, which are accompanied by a displacement of air that destroys houses and buildings and by radioactive effects that last for years.

A nightmare scenario, which does not let the security services of countries around the world, threatened above all by international terrorism, sleep. A threat, the atomic one, which until now was kept under control by technological factors, since the availability and management of a nuclear device is not within everyone's reach.

But if a portable atomic bomb existed, the international security scenario would radically change. What would happen if a phone call announced that there is a bomb triggered in the center of Rome or Paris, hidden in some waste bin? Or what if some fanatic kamikaze decides to blow himself up with a nuclear device in his pocket in the middle of a crowded stadium?

For rich terrorist groups it will be very difficult to resist the temptation to have such a powerful weapon.

But does the portable atomic bomb really exist and what are the risks of it falling into the wrong hands?

Let's take a step back in time for over half a century.

We are in the year 1950 and at the University of Berkley, in California, a new transuranic, synthetic and radioactive element is synthesized: californium. Named in honor of the California state and the University in Berkley, nicknamed "Cal".

Californium is one of the very few transuranic metallic elements to have practical applications, such as used to start nuclear reactors and in the treatment of cancer.

It has a fairly common appearance for a metal, is silvery in color and melts at around 900 °C, but it is so dangerous that a huge protective barrel weighing more than 50 tons is required to carry one gram of californium.

One of its isotopes, Californium-252 is a very powerful neutron emitter, which makes it extremely radioactive and dangerous. A single microgram of californium-252 emits 2.3 million neurons per second, resulting from spontaneous fission.

As if that weren't enough, Californium-251 also has a very small critical mass, around 5 kilograms, making it ideal for the production of an atomic bomb weighing only 2 kilograms.

But the possibilities for such a powerful pocket weapon would be endless. A californium bullet, for example, fired from an ordinary pistol, would have a destructive impact equal to about ten tons of TNT. Not to mention californium hand grenades, which in a war scenario could quickly and dramatically change the fate of a conflict.

For some, the pocket atomic bomb is just an urban legend, as the costs to produce a californium bomb would be prohibitive, around $ 10 million per gram and possibly more.

But according to a Russian general, Aleksandr Lebed, there are about a hundred small pocket atomic bombs scattered throughout the territory of the former Soviet Union, from Ukraine to the Baltic countries. Weapons supplied to special brigades of the secret service of the general staff of the USSR and the ideal tools for nuclear terrorism, easy to transport and operated by a single person.

On the other hand, the production of portable nuclear weapons, capable of being placed by a single soldier behind enemy lines, was started in the USSR since the seventies.

The recent dramatic events in Paris, where Islamic terrorists have sowed death and terror with a Kalashnikov, could be a joke compared to what could happen if the californium bomb gets into the possession of Islamic extremist groups.

It is no coincidence that our journey into the world of rare metals started right from Californium: totally unknown, with frightening but potentially extraordinary and incredibly expensive properties. Characteristics that, as we will see, are common to almost all rare metals.

100 YEARS BETWEEN WAR AND PEACE

The most informed people associate the word tungsten with the image of an incandescent light bulb, whose filaments are made of tungsten. However, the importance of this metal lies elsewhere, having accompanied much of the technological advances of humanity over the last 100 years.

There are very few people who can say they know tungsten, if not in a vague and indeterminate way.

The belief of most people is that tungsten is mainly used in the filaments of incandescent light bulbs. In reality, many of the things that are most dear to us and that we consider indispensable for our happiness, depend on tungsten. Is this an exaggeration? By continuing to read, you will be convinced of the opposite.

Tungsten is an irreplaceable metal in times of war, important in times of peace and extremely difficult to produce. Tungsten as a metal does not exist in nature and, like many other rare metals, it has fascinating and surprising properties.

Tungsten is an irreplaceable metal in times of war, important in times of peace and extremely difficult to produce. Tungsten as a metal does not exist in nature and, like many other rare metals, it has fascinating and surprising properties.

It is the hardest metal ever used by man (in hardness it is second only to diamond) and has the highest melting point of all elements: 3422 °C. These two characteristics make it indispensable for our society and the role that this metal has played in the past is only the prologue of the role it will assume in the near future.

The importance of tungsten emerged for the first time during the First World War, a period during which it assumed the status of strategic metal. According to some British observers of the time, Germany was unable to produce enough ammunition to support its military commitment and this could have had adverse consequences for the German empire. But surprisingly, it turned out that the Germans had increased ammunition production and would soon overtake the Allied arsenals.

It took hard intelligence work to uncover the secret of Germany's new manufacturing technology: high-speed tungsten steel, which made it possible to employ cutting tools to more efficiently produce armaments for their troops.

On the eve of the Second World War, in 1936, Adolf Hitler struck a trade agreement with China to secure 45% of the tungsten extracted. Thanks to this agreement and the new armor-piercing tungsten shells, Germany achieved rapid success during the campaign in North Africa. British tanks were easily pierced by German bullets, and troops called Tiger Tanks gained legendary fame.

In our day, tungsten armor-piercing bullets have been replaced by the infamous depleted uranium bullets, known for the high number of cancers and other serious syndromes among the military who participated in campaigns in Iraq, Somalia, Bosnia and Kosovo.

The Pentagon is currently seeking to increase its strategic tungsten stocks in order to cope with periods of metal shortages without jeopardizing national security.

But even moving away from the military world, we discover that tungsten is indispensable for many products of our electronic age.

It is vital for the production of numerous metal alloys and essential for increasing the efficiency of important tools for mechanical processing. Most of the processes that make it possible to obtain minimum tolerances, therefore great precision, are made possible by steels alloyed with tungsten.

In addition to its use in filaments for incandescent and fluorescent light bulbs, tungsten is essential for X-ray machinery, camera lenses, cars, airplanes, telephones, radars. But this list includes only a small fraction of the fields of application of this versatile metal.

It was recently discovered that tungsten immersed in an atmosphere of liquid nitrogen together with high voltage electrical discharges, thanks to still unknown physical processes, sharpens up to the thinnest tip in existence: an atom!

Even if tungsten metallurgy is in its infancy and there is no way to estimate how much our needs will grow in the future, it is highly likely that it will be increasingly indispensable for our technological development.

Unfortunately, to date, the availability of this metal is quite critical, as evidenced by the British Geological Society. For years, China has provided about 85% of the total world supply of tungsten. But in recent years, Beijing has begun a policy of quoting the metal externally, limiting exports, to favor supply to its industries. In addition, many Chinese mines are closing, to favor large mines.

In short, if gold remains the symbol of wealth, tungsten represents an emblematic and fascinating case, as a symbol of technological progress and sustainable development that await us in the near future.

THE WAR FOR THE 75TH ELEMENT

With our eyes on the emergency plans drawn up by the United States, should an attack on nuclear infrastructure in North Korea become necessary, the threat of another war to control the resources needed by military systems does not currently seem likely.

Yet it is clear to everyone that, long before wars fought on the battlefield, wars are won or lost in R&D and industrial production lines for defense or in developing new weapons. In some cases, decisions made years before a conflict result in the supply or shortage of the raw materials necessary for victory.

These considerations are by no means new to military experts, who have long understood the need to have immediate access to the raw materials needed for the military.

It is precisely for this reason that for most of the 20th century, the United States has maintained a strategic supply of materials, in substantial quantities: rubber, tin and other basic consumables needed to power the military machine.

The World Wars and the Cold War are over, but the imperative of having the availability of strategic military resources, even in the 21st century, is clear, or should be, for those involved in national military defense. Today the key components for the high technology of defense systems are elements of the periodic table with exotic names and unknown to the layman.

Rhenium is a striking and emblematic case.

Rhenium, a metal unknown to most people, plays a central role in the

Pentagon's defense plans and a supply crisis could have unpredictable and dramatic consequences.

This element, whose atomic number is 75, was only discovered in 1925 by some German researchers (the name derives from the Rhine river). Until 20 years ago it was a laboratory curiosity, but has since taken hold in specialized uses. The quantity produced every year in the world, about 40 tons, only makes us guess how precious this metal is and how much it could become even more so.

In the traditional economy, rhenium is used to make unleaded gasoline, in liquid gases and jet engines, for example in the Boeing 777. In the homeland security sector, rhenium is used in small rockets for repositioning. satellites in orbit, as super-alloy in high-performance engines for military aircraft such as the F-15, F-16, F-18, F-22 Raptor and the brand new Fighter F-35 Joint, which went into production in 2010, in addition to the Stealth bomber, the aircraft invisible to radar.

Aerospace engineers choose this metal for its ability to maintain strength, shape and conductive properties even at extremely high temperatures.

Rhenium is not mined, but is recovered as a by-product during the processing of copper and molybdenum. Special purifiers capture the rhenium particles in the dust released in the suction ducts.

About 14% of rhenium is produced in the United States, the remaining 86% comes mainly from Chile and Kazakhstan. For the United States, this dependence on foreign suppliers constitutes a vulnerability, which in the event of interruptions, whether accidental or intentional, would put the entire national security at risk.

Think of it as the other side of the coin of economic globalization: supply chains unfold all over the planet, but they can be interrupted suddenly and without warning.

How to ensure that rhenium is available, especially for military defense needs? The US government has strategic warehouses where critical materials can be stored for use in times of need. But it could also encourage allied countries to recover rhenium from all their copper and molybdenum production plants.

But still few among those responsible for national defense are aware of the importance of strategic stocks of scarce but indispensable elements for our safety such as rhenium and about a third of other elements in the periodic table. Without this awareness, we may find that all high-tech war machines can be brought to their knees by the lack of a handful of rhenium.

THE SECRET METAL OF SOVIET AVIATION

In the middle of the Cold War, Soviet engineers built a fighter plane that terrorized the Pentagon. To thwart the threat, American agents attempt an impossible mission: to steal the airplane prototype. Fantasies of a novelist or reality?

In the mid-1970s, a book was published, Firefox, which narrated the mission of an American agent in the Soviet Union with the aim of stealing a brand new fighter jet that had amazing performance and capable of threatening the Western military blockade.

The novel, fruit of Craig Thomas' narrative vein, was inspired by a real concern on the part of Americans who, in the middle of the Cold War, had lost their technical superiority in some areas of aerospace design.

Soviet engineers had managed to design and build two fighter aircraft, the MIG-21 and the MIG-23, with amazing characteristics for the skills of the time and to do so they used a secret metal: scandium.

Scandium is the eighth rarest element on Earth and is a powerful particle refiner. If added to aluminum alloys it increases their resistance and durability by 50%. To better understand what this means, just think that if a carbon structure were to offer the same performance as a scandium alloy, it would weigh much more than a scandium structure.

Scandium, the lightest of the transition metals, also increases stiffness and stress resistance, improves quality, durability, and inhibits recrystallization of the aluminum alloy.

Dmitri Mendeleev, the inventor of the periodic table, predicted the existence of scandium before it was discovered in nature. The discovery took place only in 1879, thanks to a chemistry professor who did not find a better idea to call the new element after his homeland, Scandinavia. This rare metal was isolated in its purest form only in 1937 and the first pound of pure metal was produced as far back as 1960.

Since scandium has some characteristics similar to those of elements such as yttrium and lanthanum, it is often classified as a rare earth. However, scandium has a low affinity for other minerals and is rarely found in concentrations that make a commercially usable deposit.

The quantities of scandium available are scarce and expensive (about $ 5,000 per kilogram) and this is a major problem, especially for the military, the main users of the metal.

Scandium currently has three types of uses: to strengthen aluminum alloys and therefore in the aerospace industry, to build high-intensity lamps (used in television studios) and light bulbs, as a radioactive tracer in oil refineries.

In recent years, the use of scandium, alloyed with aluminum, for the construction of sports equipment such as bicycles and baseball bats has become increasingly popular.

The world's largest scandium deposit was recently discovered in Australia during excavations at the former Greenvale nickel mine in northern Queensland.

Now that the Pentagon has discovered the secret metal of Soviet military aviation, all that remains is to accomplish the most challenging and difficult mission: to buy scandium at reasonable prices.

THE BLACK PRINCE OF ALL METALS

Fascinating for its surprising liquid characteristics but equally dangerous for its toxicity, mercury is surrounded by ancient legends that identify it as the origin of all metals.

Mercury (hydrargyrum in Latin) is in fact considered the cursed prince of all known metals, fascinating but deadly.

According to the ancients, it was the primordial substance from which all other metals were born and the alchemists believed that, by changing the sulfur content, it could be transformed into any other metal, including gold. However, its dangerousness is such that there is an international treaty to contain its use.

The charm of this metal also derives from the fact that it is the only one that is liquid at room temperature and one of the rare substances that reacts with the noblest of metals: gold. The process by which the reaction between the two metals takes place is simply extraordinary to see: a gold leaf that comes into contact with mercury first melts and then dissolves completely. According to the alchemists, at this point in the process, by evaporating the mercury, pure gold could have been obtained.

But there is also another face of mercury, damned dangerous. It is a deadly poison and with long-term effects for all human beings, but also for other living organisms. Some historians believe that Napoleon, Ivan the Terrible and Charles II of England died poisoned by this metal.

More than a third of the mercury released into the environment is a direct consequence of our craving for gold. Worldwide, it is estimated that 10 to 15 million small-scale miners mine and dredge gold using mercury to separate the metal.

In water, this metal transforms into a highly toxic organic molecule, methylmercury, which is absorbed by algae and plankton. These are eaten by larger animals, which are in turn eaten by larger creatures, which in turn are often eaten by humans. During the entire process, mercury becomes increasingly concentrated and poses a serious threat especially to the developing brain of babies and fetuses in the womb.

The governments of the world are not very much in agreement on what to do to deal with the threat. However, in 2013, 93 countries, including Italy, signed the Minamata treaty to reduce mercury pollution.

Minamata is the Japanese city that gave its name to the neurological syndrome caused by mercury poisoning, which occurred in the mid-21st century, when the contaminated waters of a chemical industry arrived in the food chain through the fish, shellfish and crustaceans of the homonymous bay.

THE ISOTOPE THAT IS DISAPPEARING

There are radioactive isotopes on which the health of many people, in the present and in the future, depends. The production of one of these risks disappearing forever.

When most people hear about radioactive isotopes, they start to worry. In the case of molybdenum-99 the concern is entirely justified, but not for its dangerousness but rather for its lack.

Someone knows molybdenum as a rare metal, used in many industrial applications: an additive to produce steel, indispensable in many applications in the aerospace and electronics sectors.

But molybdenum-99 is quite another thing. In fact it is one of the 35 known isotopes of molybdenum, whose decay produces a radioactive isotope called technetium-99, a key component of nuclear medicine, that branch of medicine that uses radiation to collect information on the internal organs of the human body, generally to diagnose the diseases.

Technetium-99 can be used to examine a wide variety of organs and tissues and for this reason it is the most commonly used radioisotope in medical diagnosis. Currently, around the world, 40 million diagnoses are made per year and technetium-99 accounts for 80% of all nuclear medicine procedures.

It is clear that technetium-99, and by extension molybdenum-99, is a very important product. Unfortunately, in recent years, the delicate balance between the supply and demand of molybdenum-99 has been broken.

The entire production is based on 5 very old nuclear reactors, which have been in operation for 50 years, well beyond their limit, since they should have lasted no more than 30 years.

When one of these reactors, the HFR located in the Netherlands, stopped for two months for routine maintenance, the market lacked 33% of the molybdenum-99 needed.

But the age of the reactors is not the only problem plaguing the supplies of molybdenum-99. The American Nuclear Security Administration is very concerned that most molybdenum-99 is produced by processes that require highly enriched uranium (HEU), which is the material used for the production of nuclear weapons.

In other words, the technology currently used to produce molybdenum-99, and consequently technetium-99, is obsolete and potentially dangerous. A better and safer technology would be that with low enriched uranium (LEU).

Unfortunately, at the moment, the companies that are trying to develop new procedures for the production of molybdenum-99 are few and of small size, engaged in a company with high costs and uncertain times.

This is why the worldwide stability of molybdenum-99 is likely to falter and cause serious problems for the entire global diagnostic sector.

A 548 MILLION DOLLAR TREASURE DISCOVERED BY ACCIDENT

The mining history of Peru and of an important metal for the entire world industry is mainly linked to chance and cold.

In fact, it seems that the largest vanadium deposit in the country, Minas Ragra, was discovered in 1905 by some mining workers who, during a break from work, went to look for coal to warm up due to the cold temperatures of the mountains. Unbeknownst to them, they collected vanadium and burned it, breathing in toxic fumes that nearly killed them.

The Minas Ragra mine was officially discovered shortly after and purchased by some American investors, who sparked a vanadium rush across the South American country. In 1920 it was already clear the vastness and wealth of the mine, which had become indispensable for the development of the very young and thriving American automobile industry.

But the mine would soon run out. After producing 52% of all vanadium in the world, dropping the price from $ 2,260 per kilogram to a low of $ 2.2 per kilogram, sunset was approaching for one of the first industrial-scale vanadium mines.

By 1955 Minas Ragra had run out of even the last gram of vanadium and the world was left without 43,000 tons of vanadium pentoxide. According to some estimates, the Peruvian mine had produced about $ 548 million of vanadium up to that point, at current market prices.

For more than 60 years, all ambitions to explore new mines in the country disappeared and no business was started to extract vanadium from Peruvian soil.

Recently, interest in restarting some vanadium mines in Peru has reignited, thanks to neighboring Brazil, where vanadium production in the Maracas Menchen mine, managed by Largo Resources, is steadily increasing, from which they have already been extracted. about 1,140 tons.

However, the exhaustion of this mine too could put into play a country like Peru which, as some old miner claims, still hides rich and unknown mining treasures in its mountains.

THE SCARY METAL

It is one of the most important metals for our society and will become even more so in the near future. Better to know it than to be afraid of it ...

We are talking about uranium, a name that hardly does not instinctively inspire a certain discomfort, mixed with fear, generated by everything we have heard about the dramatic nuclear accidents at Chernobyl and Fukushima and about the use of depleted uranium in war zones.

The mass media, on the subject, have never been read and, to give some more emotion to their audience, they preferred to terrify rather than inform.

In reality, uranium is a fairly common metal, present in most rocks in low concentrations (2 to 4 parts per million). Moreover, in recent years it has attracted the attention of many investors, who see the possibility that the prices of the metal could put a sign of substantial gains in the near future.

Uranium was discovered in 1789 in the pitchblende mineral of Martin Klaproth, a German chemist, who baptized it as the planet Uranus.

When refined it is a silvery white metal, weakly radioactive, but which reacts with most non-metallic elements and their compounds, which increases with temperature.

Uranium is present in the form of two isotopes (atoms with one more or less neutron): uranium-238 (U-238) and uranium-235 (U-235). The first represents more than 99% of the available metal, the second less than 1%. The rarest, U-235, is also the most important and is commonly used as a nuclear fuel. In fact, it is fissile, which means that under certain conditions the isotope can be divided, releasing a significant amount of energy.

U-238, on the other hand, is not fissile but fertile. What does it mean? It means it can capture one of the neutrons around a reactor core, creating plutonium-239, a fissile isotope that gives off a significant amount of energy. Plutonium is infamous for being used in the atomic bomb that was dropped on Nagasaki (the Hiroshima one was uranium-235).

Currently, the most important use of uranium is in the production of nuclear energy. It was used in the first nuclear power plants in 1950 and, to date, nuclear reactors have become more than 400, a fleet of power plants that provides over 10% of the world's electricity.

But, as everyone knows, there is a somewhat less peaceful use of uranium: high-density penetrators and nuclear bombs.

The former are depleted uranium ammunition bound with one or two percent other metals, usually titanium and molybdenum, while nuclear bombs have dramatically constituted one of the first uses of uranium even though, since 1990, most of the military uranium has been reconverted for use as fuel in civilian nuclear power plants.

With the population of our planet continuously growing, the need for energy sources is more important than ever. By 2030, electricity consumption is expected to double from 2007 levels and a significant portion will come from nuclear energy. China alone will build 40 new nuclear reactors by 2020, as will Russia which will build another 25 and India another 24.

So it is questionable whether there will be enough uranium to meet all these new needs. According to many analysts, there is no doubt that there will be a shortfall in the supply of this metal and with it a sharp increase in prices. But the timing with which this will happen is not too clear, as evidenced by the fact that for some years now, experts have been expecting a rise in prices that has not been there until now.

However, for those who believe that the fundamentals of supply and demand are the most important drivers of the market, there is little doubt that uranium is an interesting investment for the next few years.

THE NUCLEAR FUEL OF THE FUTURE

The resurgence of nuclear power in the post-Chernobyl era has long been blocked by the high cost of new nuclear power plants and the lifespan of much of the radioactive nuclear waste, which can extend well beyond 10,000 years.

But a growing number of scientists believe that an alternative nuclear fuel to uranium and plutonium could solve the problem. The alternative metal is called thorium and could pave the way for the production of cheaper and safer nuclear power.

Thorium is a weakly radioactive metal that was discovered in 1828 by the Swedish chemist Jöns Jakob Berzelius, who named it in honor of Thor, the god of thunder. Thorium is found in small quantities in most rocks and soils, where it is about ten times more abundant than uranium and is about as common as lead.

Thorium is an energy bomb: one ton of it can generate the same energy as 200 tons of uranium.

In the 1950s, some American physicists had considered thorium as a source of energy for nuclear development and in 1957 the Shippingport plant was inaugurated, a small plant of just 60 Megawatts of power, totally powered by thorium.

But uranium, which has plutonium as a byproduct, took hold in the then nascent nuclear technology thanks to the latter's military uses. In fact, plutonium was the most used element in weapons produced during the Cold War. It is not possible to extract plutonium from thorium and consequently it is impossible to produce nuclear weapons.

A Japanese company working on thorium-powered molten salt reactors estimates that the power generated by such a reactor would cost at least 30% less than the energy produced by today's light-water reactors. In addition, molten salt reactors could burn stocks of hazardous waste produced by previous generations of nuclear reactors.

Only India has focused on this technology which has come back into fashion in recent years. The United States and especially China are beginning to invest resources in the development of nuclear power plants fueled with thorium. It seems that China will inaugurate its first plant later this year.

Thorium is also present in Italy in fair quantities, in Lazio, on the border between the Aosta Valley and Switzerland and on Etna. According to Carlo Rubbia, Nobel Prize for Physics, there are also deposits in Umbria and Abruzzo.

Few people know that in 2000, in Italy, Enea began work on the Rubbiatron, an energy-amplifying nuclear reactor flanked by an external proton source, with thorium rods as a fissile material and liquid lead as a coolant. Born from an idea of Carlo Rubbia, this reactor would be achievable with current technologies and would have undoubted advantages over even the traditional latest generation reactors. However, the construction by Enea was abandoned due to lack of funds.

OR RARE EARTH OR LIFE

In 2014 a violent regime purge in North Korea marks the beginning of a maxi-operation for the exploitation of the largest rare earth deposit in the world.

The rare earth market had just gotten into turmoil with the announcement of the discovery of the world's largest deposit in North Korea, with a potential of 6 billion tons of ore, with an estimated value of $ 65 trillion.

But a week after the announcement of the discovery and licensing of Pacific Century Rare Earth Minerals Ltd, something dramatic happened: the execution of Jang Sung-taek, uncle of Kim Jong-un, North Korea's supreme leader.

Jang was found guilty of "various debauchery" and more precisely "of having led a capitalist lifestyle aimed at dragging the country into decadence through the distribution of all kinds of pornographic images, leading a dissolute and depraved life, with waste of money wherever he went ". And yet he is called "traitor to the nation", "worse than a dog" and "despicable human scum", terms that are usually reserved for the leaders of South Korea.

But among the charges that led to the death sentence of poor Jang Sung-taek and the killing of all his family members, there is a sentence that sent chills to all the executives of Pacific Century Rare Earth Minerals. Ltd: "Jang Sung-taek has sold the country's precious resources at rock bottom prices."

There are probably many factors behind Jang's execution, including the new discovery of rare earths. Certainly, the story of the execution of the dictator's uncle has triggered concerns about the possibility of exploiting the new field by relying on foreign investments.

According to Leonid Petrov, a Korean researcher at the Australian National University's College of Asia and the Pacific, Jang's death demonstrates that North Korea is resistant to change and has no interest in making the reforms that are needed to support foreign investment in the country's economy. After all, crisis and isolation are two necessary conditions for maintaining the regime.

According to a White House adviser, the episode will have a series of ripple effects including the control of the extraction of rare earths in the country. Of course, Pacific Century Rare Earth Minerals Ltd has learned the hard way what the concept of political risk means when investing in North Korea.

THE 10 MOST RARE METALS IN THE WORLD

Someone thinks that gold is the rarest metal in the world, due to its preciousness. In reality, there are much rarer and also much, much more precious metals.

If you found yourself the emperor of a powerful kingdom and you had to choose to mint national coins with a metal that is impossible to counterfeit, thanks to its scarcity and rarity, you could choose from those contained in the ranking of the 10 rarest metals in the world.

Rare does not always rhymes with precious, since the value of a metal is determined not only by its scarcity, but also by market demand.

In short, if the supply of a metal is low but the demand is also low, its value may not be high. This is the case of the rarest metal in the world, iridium, which has modest prices and very limited uses.

IRIDIUM - It is the rarest element on the entire earth's crust (0.0004 parts per million), about 12 times rarer than gold. According to some important scientific studies, the origin of the metal is extraterrestrial, arrived with the same asteroid that led to the extinction of the dinosaurs, crashed near the current Yucatan peninsula (Chicxulub crater).

RHODIUM - Like iridium, it belongs to the platinum group and for some years was the most precious of metals. Extracting rhodium is a rather complex undertaking, in fact this metal is found mixed with minerals of other metals, such as palladium, silver, platinum and gold. Even the smelting operations are very difficult, so much so that a total world production of only 7 tons per year is barely achieved.

• TELLURIUM - In the molten state, tellurium is able to corrode metals such as copper, iron and even stainless steel.

- RUTHENIUM - It is a very difficult metal to produce due to its particular chemical-physical characteristics. Therefore ruthenium is commercially available in small quantities and its prices are particularly high (currently around $ 60 per ounce).

- OSMIUM - Osmium also belongs to the platinum group and its tetroxide is used for the relief of fingerprints. It is the heaviest metal in nature.

- RHENIUM - Rhenium does not exist in nature in the free state and not even in the most common minerals. The only possibility is to obtain it from ammonium perrenate.

- GOLD - Gold is distributed a little over the entire earth's crust, with an average concentration of 0.03 parts per million, corresponding to 0.03 grams per ton.

- PLATINUM - It is present in nature in its pure state or in an alloy with iridium. Its highly toxic compounds are quite rare in nature and some of them, for example cisplatin, are used as anti-cancer drugs.

- PALLADIUM - Also belonging to the platinum group, it is widely used as a catalyst.

- BISMUTO - It is mainly used in the pharmaceutical sector and for the preparation of low melting point alloys such as, for example, those for fuses.

RARE METALS, MINOR METALS OR TECHNOLOGICAL METALS?

TECHNOLOGICAL METALS

From a curiosity for scientists only at the beginning of the last century, to becoming the secret elements to win the Second World War. The history of technological metals is full of surprises that reach up to today.

The term technological metals is relatively recent and was introduced, or rather reintroduced, by Jack Lifton in 2007.

We can say that technological metals are those metals, generally rare metals, which are essential for the production of many high tech devices, engineered systems, armaments, medical devices, telecommunications devices, such as:

- mass production of miniaturized electronic devices;

- advanced weapons and national defense platforms;

- the production of electricity with alternative sources, such as solar panels and wind turbines;

- the storage of electricity from cells and batteries.

There are of course numerous other applications of these metals.

Almost all technological metals are by-products of the production of common metals, with the exception of rare earths and lithium.

Before the Second World War, there were many metals for which there were no practical uses. They were literally a laboratory curiosity and were only available in small quantities, obtained at high cost in terms of time and money. For this reason, they were called the minor metals, simply because they had no practical uses unlike the common metals and precious metals. It was not important how abundant the metal was in nature, but only if it had a practical use or not (also because the quantities produced were linked to this consideration). Nickel, for example, was a minor metal before the commercial development of stainless steel in 1919, when the mass production and use of stainless steel became predominant. Nickel became a high-production metal and is now classified as a common metal.

In the early twentieth century, malleable tungsten was developed by General Electric and became a material widely used in incandescent light bulb filaments. Shortly thereafter, tungsten steels were developed and used, initially for armor and armor-piercing shells for military use. Later, tungsten carbide was used in cutting tools and constituted a revolution in precision machining, just in time to turn engine production into mass production. Tungsten, a minor metal in 1900, became an important industrial metal in 1918 and was already referred to as a technological metal.

But the clearest example of a metal that has passed from a minor metal to a common metal is aluminum. At the end of the nineteenth century, aluminum was a minor metal. It was used to cover the Washington Monument in 1886, as a symbol of America's wealth. Aluminum was therefore more expensive than gold. Only a madman or a visionary could have predicted in 1886 that ordinary people would cook with aluminum pots and pans only a century later. Even in 1919 the idea o f stainless steel appliances for ordinary people would have been considered a fantasy.

World War II transformed a dormant academic discipline, the study of the physical properties of metals, into modern metallurgy seeking to develop new uses of metals and implementing new products based not only on their properties as structural materials, but more importantly, on their properties. their new electrical, electronic and magnetic properties for use in modern technologies.

Fifty years ago it was not clear that if some minor metal would be used in the production of mass goods. We were discovering that the electrical and magnetic properties of the chemical elements were able to satisfy the needs and desires of our civilization. Until the First World War, metallurgy knew only the structural, decorative and electricity transmission properties of metals. The last metal discovered, rhenium, dates back to 1924. What no one knew, in the period between the two world wars, was the importance of knowing which, among the minor metals, could have been produced in significant volumes to follow the growth of mass production.

There was no need to know, especially in the academic world, where most of the studies on these metals were conducted. The equation was simple: no use equals no demand and therefore no effort was needed to supply these metals in large quantities.

The Second World War was the most important event for the transformation of minor metals into technological metals. The economic problems that limited innovation were put aside and national security (i.e. winning the war) became the sole driver of technological development for jet engines, radios, radar, electronics, information technology and super-guns.

A fantastic galaxy of the best physicists, engineers and chemists was gathered from the various world governments, a collection of intelligence that occurs perhaps once every thousand years. The metals that were deemed necessary were made available without any financial constraints. Chemical engineers began to learn how to find, refine and produce in large quantities metals hitherto considered minor, to meet the exaggerated technological demands of the ongoing war. Among other things, ultra-pure silicon and germanium, gallium and indium, uranium and thorium, rare earths and, immediately after the war, lithium were produced in quantities never seen before.

After the Second World War, 50 years of the Cold War began, during which the anti-economic and mass production of minor metals for military use continued. The surplus production was diverted to civil applications, in search of economic and mass uses. These events were the germs of the current "technological age". The economic considerations were very simple: the minor metals are used for war, hot or cold, and the states totally subsidize their development and production.

What are technological metals and what are their main uses? What is the difference between a rare metal and a technological metal?

Today we are totally dependent on the technological metals that are needed in the production of mass consumer goods such as electronic devices, televisions, cell phones, computers and all communication devices. Our life depends on technological metals and the very concept of Homeland Security is linked to these metals as regards advanced weapons and telecommunication systems.

But what are the technological metals? Here is a list produced by the US Geological Survey and the British Geological Survey, with the estimated world production in 2009 (tech metals are in bold while rare metals are underlined):

- Cobalt 62,000 (tons)
- Uranium 35,332 (tons)
- Lanthanum 32,860 (tons)
- Silver 21,332 (tons)
- Neodymium 19,096 (tons)
- Cadmium 18,000 (tons)
- Lithium 18,000 (tons)
- Yttrium 8900 (tons)
- Bismuth 7300 (tons)
- Praseodymium 6150 (tons)
- Gold 2350 (tons)
- Dysprosium 2000 (tons)
- Selenium 1500 (tons)
- Samarium 1364 (tons)
- Zirconium 1230 (tons)
- Gadolinium 744 (tons)

- Indium 600 (tons)
- Terbium 450 (tons)
- Europium 272 (tons)
- Palladium 195 (tons)
- Platinum 178 (tons)
- Germanium 140 (tons)
- Gallium 78 (tons)
- Rhenium 52 (tons)
- Rhodium 30 (tons)
- Hafnium 25 (tons)
- Tantalum (?)
- Erbium (?)
- Holmium (?)
- Lutetium (?)
- Scandium (?)
- Tellurium (?)
- Thorium (?)
- Thulium (?)
- Ytterbium (?)

Technological metals are almost all also rare metals and are often obtained as by-products of the common metals.

The problem with technological metals is that their supply, or rather our maximum production rates, largely depends on the production of common metals. In the case of rare earths, the main problem lies in the complexity of the metallurgical process for the separation of the individual metals.

Rare earths and lithium are currently the subject of much discussion, as they have become highly visible technological metals.

The definition of rare metal is quite fluid, some rare metals to date have not always been such. Lithium, for example, is on the verge of entering the list of rare metals, due to its use in electronic memories.

But it is historically proven that once a minor metal becomes a technological metal, it will never go back to being a common minor metal.

RARE METALS

Rare Metals is the conventional name given to a group of over 50 metals, some of which are listed below.

- lithium, rubidium, cesium, beryl (LIGHT)

- titanium, zirconium, hafnium, vanadium, niobium, tantalum, molybdenum, tungsten (TRANSITIONAL)

- gallium, indium, thallium, germanium, selenium, tellurium, rhenium (POST-TRANSITIONAL)

- scandium, yttrium, lanthanum and lanthanides (RARE EARTHS)

- francium, radium, actinium, thorium, protactinium, uranium, plutonium and other trans-uranic elements, polonium, technetium (RADIOACTIVES)

They are also referred to as technology metals or strategic metals or minor metals. Some define such metals as rare when annual world production is less than 25,000 tons.

They are relatively new metals for technological applications or, some of them, have found limited practical applications to date. But the production and fields of application of these metals are continuing to expand. The term rare metals came into use in the USSR in 1920 and these elements are sometimes referred to as less common metals. The rarer metals are often dispersed in the earth's crust; this together with the considerable technological difficulties encountered for extraction and refining, explain why they were discovered relatively late.

The production of rare metals has been developing at a particularly high rate from the Second World War to the present. They are essential metals for almost all high-tech sectors: aviation, missile, electronics, energy and nuclear engineering.

Among the numerous applications are:

- the production of electronic devices;
- advanced weapon systems;
- solar panels and wind turbines;
- conservation of electricity with batteries and cells.

Of course, with the increase in demand and use of these metals, the denomination rare metals tends to lose its original meaning.

Rare metals are usually present in small concentrations in minerals. The chemical processes of isolation, separation and purification are fundamental for obtaining rare metals from minerals and the technology allows more or less efficient processes. Many rare metals are contained, in small parts, even in common metals.

According to many observers, rare metals will replace oil in our century, also in view of the imminent transport revolution linked to the development of new automotive electric technologies.

RARE EARTH ELEMENTS

17 EXOTIC ELEMENTS

Europium, samarium, lanthanum, are nothing more than the names of some of the seventeen minerals that are part of the rare earths (belonging to rare metals, or minor metals) and are represented in the periodic table as chemical elements.

They are materials with unusual names, widespread almost everywhere in the earth's crust, whose extraction involves techniques that are not too different from traditional ones, but whose extraction produces a high rate of pollution.

Without these rare elements it would not be possible to produce anything of what today is the most advanced industry.

Neodymium, for example, is the essential element for the production of batteries and engines of hybrid or electric cars, for computer hardware, for mobile phones and for cameras. In the military field, neodymium oxide is an indispensable ingredient in the magnets that operate the directional wings of precision missiles. Europium and yttrium are used to produce optical fibers and green light bulbs, while scandium is the raw material for large lighting systems in sports stadiums.

In the early 1990s Deng Xiaoping had proclaimed that "rare earths are for China what oil is for the Middle East" and currently none of the large multinationals, from Philips to Siemens, from Toyota to Nokia, from Hewlett Packard to Apple, up to Sony and Canon, can manufacture their own devices without sourcing from China.

Estimates say that 12% of the fields are in the United States, 18% in the former Soviet Union, smaller quantities are scattered in many other countries and, according to estimates, between 37% and 58% reside in China. We also find many mines in Afghanistan, but the costs of extracting rare earths are very expensive and not competitive with the Chinese one, which sells them all over the world at a very low price.

But since 2009 China has drastically reduced exports of rare earths, saying it must preserve them for environmental reasons and for its own needs.

A situation that worries high-tech industries, especially Japan, to which Beijing has even blocked exports during a dispute over sovereignty over a group of islands. Now Tokyo plans to recycle rare earths, as well as to seek substitutes for them.

The United States, Australia and other producers had stopped the extraction because it was not profitable, in the face of cheap Chinese production. But now the research and extraction of these minerals has resumed, although it will take time to achieve adequate production. And Beijing shows all its intentions to leverage its market power in this field, to force the rest of the world to accept its conditions: these involve not only a net transfer of capital, but also of work and above all of industrial secrets from West to the People's Republic of China.

China has made enormous efforts to build a strategic reserve of rare earths. Details of the storage site are not known but, according to reports from Chinese state agencies and statements from state-owned companies and state media reports, it appears that the complex was built in a region of Mongolia. With a storage capacity of rare earths that amounts to more than the total of China's exports last year (39,813 tons), the reserve could have the ability to influence the entire global market, already largely dominated by China, which, at the nowadays, it controls more than 90% of global rare earth production.

For all these reasons, many investors see the sector as rich in opportunities. However, the biggest obstacle for those who want to invest in rare earths is in finding reliable and accurate information on the subject. A first little help comes from starting to distinguish rare earths, or RE (Rare Earths), or REE (Rare Earth Elements) or REM (Rare Earth Metals), into two main categories: heavy rare earths and light rare earths.

Heavy rare earths (HREE - heavy rare earth elements):

- Yttrium, used for screens, alloys and TVs.

- Terbium, used for lasers, alloys and fuel cells.

- Dysprosium, used for lasers and TVs.

- Holmium, used for lasers.

- Erbium, used for lasers and vanadium steels.

- Thulium, used as an X-ray source and for ceramics.

- Ytterbium, used for infrared lasers and high reactivity glass.

- Lutetium, used for PET scanners and catalysts.

Light rare earths (LREE - light rare earth elements):

- Samarium, used for magnets, lasers and lights.

- Neodymium, used for magnets.

- Lanthanum, used for rechargeable batteries.

- Cerium, used for batteries, catalysts, glass and steel production

- Praseodymium, used for magnets and for coloring glass.

- Scandium, used for aluminum alloys and in the aerospace industry.

- Europium, used for TV screens.

- Gadolinium, used for magnets and superconductors.

- Promethium, used for nuclear batteries.

That said, our journey through rare earths does not end here and, in the next chapters, we will talk about the most important of these rare elements.

RARE BUT NOT TOO MUCH

When you are in front of a glass of rum and you are about to smoke your favorite Cuban cigar you cannot do without, more or less consciously, cerium, one of the rare earth elements.

To be precise, all cigarette lighters work thanks to flints that trigger sparks. Stones that are produced using an alloy, called *mischmetal*, composed of 50% cerium, lanthanum and in small percentages of neodymium and praseodymium.

But cerium, which looks quite similar to iron, is also used in the production of aluminum alloys, magnesium alloys and in some steels.

Cerium is a bit like the black sheep of rare earths, since, among these, it is the most abundant element on the earth's crust. A metal that has been suffering for some time from an oversupply and whose prices are very low, so much so that it does not even guarantee the costs of separation and purification.

However, according to the US Department of Energy's Critical Materials Institute, the luck for this metal could come from a new type of process for the production of nylon and stabilizers for PVC (indispensable products for the production of plastic), in which palladium would be used as catalysts. and cerium.

Although the thing is in the early stages of development and, therefore, it is difficult to say whether it will affect the cerium market, the idea is very interesting.

Soon, the new process will be moved from the laboratories to the actual production, where there will be a way to measure the extent of this novelty, which should allow greater energy efficiency and a reduction in hydrogen consumption.

China is abandoning the use of lead stabilizers for the production of PVC, creating the opportunity for new stabilizers, such as those containing cerium, to succeed.

To date, the abundance of worldwide stocks of cerium, also treated as a waste material, makes it quite unlikely that the new technology will have any impact on prices in the short term. The huge stocks that have accumulated, in China and elsewhere, suggest that it will take many years before they can be disposed of.

However, it should not be forgotten that the global markets for nylon and PVC stabilizers are huge and therefore have the capacity to rapidly amplify the demand for cerium, which could happen in the next few years.

THE METAL HIDDEN IN THE ELECTRIC CAR

The heart of all electric cars beats thanks to an almost unknown metal with surprising properties: dysprosium.

It is very likely that just by reading these lines, many people have discovered for the first time the existence of dysprosium, a metal that is becoming increasingly important in the production of many high-tech devices.

Dysprosium is one of the rare earths and was discovered as far back as 1886 as an impurity. But until 1950 there was not even a sample of pure dysprosium. Its name, deriving from the Greek and meaning "difficult to reach", says a lot about its rarity.

It has a bright silvery-metallic appearance, low toxicity and no known biological role.

Like the other lanthanides, 15 metallic chemical elements with atomic numbers from 57 to 71, is found in monazite and bastnaesite deposits, as well as in minerals such as xenotime and fergusonite.

Its main use is in neodymium-based magnets, also called super-magnets. The addition of dysprosium allows the magnets to preserve the magnetism even at the highest temperatures.

Some might think that a super-magnet is a laboratory curiosity or an educational toy. In reality, this type of magnets is indispensable for the motors and generators of wind turbines and electric vehicles. But dysprosium is also used in the control rods of nuclear reactors, managing to easily absorb neutrons without swelling.

Unfortunately, dysprosium is becoming increasingly difficult to obtain and has forced some consumer goods manufacturers to reduce the quantities used. For example, in 2013, Hitachi Metals reduced the use of dysprosium in NeoMAX magnets, used in the automotive industry.

Dysprosium, one of the most expensive heavy rare earth elements, is supplied by only one country (China), which creates supply problems and high prices.

As the largest producer of rare earths in the world, it is not surprising that China is also the largest producer of dysprosium in the world. While China's tightening of the rare earth market has been weakening recently, the country still makes up the lion's share of their production.

Concerns about the possibility of a dysprosium shortage have increased in recent times, mainly due to the high demand for the magnets needed to produce batteries for hybrid and electric cars and wind turbine motors.

As the most informed investors know, a dysprosium deficit is expected for the next few years, as for all other so-called heavy rare earths. Main reason why analysts expect the price of this metal to rise.

EVERYONE WANTS IT...

There is a metal in the world whose latent demand is enormous.

It is a "dark" metal, completely unknown to most people, of which there is currently less than ten tons per year available all over the world.

The metal in question is scandium, scarce, expensive and used above all in the military field, where high performance is required.

But things are changing. As a niche metal, used in very small volumes, in the next few years we could see a much greater use of scandium. If an efficient source of scandium emerged, two huge markets would open up that could consume the metal on a large scale: solid oxide fuel cells and scandium aluminum alloys.

Over the past 50 years, dozens of patents have been filed for scandium-based materials and technologies that are just waiting for this metal to start becoming available.

Scandium, whose existence was predicted by Dmitri Mendeleev in 1860, is a soft metallic element of silver color. It is sometimes classified among the rare earths, as it is often found in the same deposits.

The applications of this metal are basically three:

- aluminum alloys,
- solid fuel cells (SOFC),
- lamps, lasers and video screens.

Scandium aluminum alloys can double or triple their tensile strength, while maintaining the same malleability so useful for producing geometrically complex elements. They also maintain weldability and corrosion resistance. Russia, during the Cold War, used alloys of this type in the production of its MIG fighter jets.

Fuel cells work by converting oxygen and a fuel source into an electrical current, water, carbon dioxide, and heat. For example, they were used by American NASA as a power source on spaceships.

The best indicator of what the scandium market will become is the yttrium market, its closest colleague in the periodic table. The two metals are similar, except for the fact that scandium is enormously more heat resistant, has greater electrical conductivity, superior optical properties and, when alloyed with aluminum, provides top-level performance.

So why isn't scandium used instead of yttrium? The answer is on the market: there are very few supplies of scandium and consequently its cost is very high, 40 times that of yttrium, even if very small quantities are required to have a dramatic transformative impact on the performance of the material where it is used.

Therefore yttrium, whose current market is about 3 billion dollars, is seriously in danger of being supplanted, at least in part, by scandium.

The fuse of this change is already lit. In fact, important scandium deposits with a degree of purity three to four times higher than those currently coming from Russian deposits have been found in New South Wales, Australia. Within two years the production of these new deposits will be on the market and the greater availability will make the costs of the metal more reasonable.

A big change, which will affect not only the yttrium market but also that of aluminum. Very soon, scandium will be a somewhat less unknown metal.

THE MAGNET THAT BREAKS THE BONES

How much strength can a magnet have? Can it be so strong that it becomes dangerous? When it comes to neodymium super magnets, there are never too many precautions...

Known to be one of the strongest magnets currently available, neodymium is a rare earth element.

Neodymium iron-boron magnets (or magnets) are used in a wide range of modern technological applications.

Neodymium was discovered in 1885 by the Austrian chemist Carl Auer von Welsbach. Its discovery sparked controversy and skepticism as to whether or not it was a real metal, since in nature it exists only as didymium (a mixture of praseodymium and neodymium). This is why it is called neodymium, from the Greek neos didymos, new twin.

It is a fairly common metal, twice as common as lead and about half as common as copper. It is mainly extracted from two minerals, monazite and bastnasite, but it can also be obtained as a by-product of nuclear fission.

Neodymium has incredible magnetic properties and is therefore used to create magnets that have enormous strength. It is usually mixed together with praseodymium and dysprosium, the latter to improve the functionality of the magnets at higher temperatures.

The force exerted by a neodymium magnet is by no means comparable to that of other types of magnet. For this reason it can be dangerous to handle this type of magnet which, just larger than a couple of cubic centimeters, are strong enough to injure parts of the body between two magnets, also causing bones to break.

These magnets are the basis of the revolution in modern technologies, such as mobile phones and computers. In fact, thanks to their power, these small-sized magnets have made the miniaturization of electronics possible.

To give some examples, the vibrations to incoming calls from phones and smartphones are produced thanks to these magnets, while MRI scanners (the so-called magnetic resonance imaging) to investigate the internal organs of the human body without using radiation, are also a application of these magnets. But they are also indispensable for the functioning of modern televisions, wind turbines and computer hard drives.

Chinese production currently accounts for 95% of the world's rare earth supply, including neodymium. For some time, China has limited the supply of rare earths to the rest of the world, raising the concerns of the international community, as the demand for these metals continues to increase. However, in 2014, the World Trade Organization (WTO) condemned China's behavior and at the end of last year the Chinese government declared that supplies would return to normal.

Many observers are skeptical that China's rare earth supply flow will resume steadily, and until mining companies outside of China succeed in developing new deposits, the supply of neodymium will remain difficult and at risk of disruption.

A scenario viewed positively by rare earth investors, since it would help price growth. Exactly the opposite of what all high-tech producers hope.

THE HUNGRY CARS OF
LANTANIUM

World demand for lanthanum is expected to rise.

Current hybrid car batteries use between 12 kg and 15 kg of lanthanum. The motor vehicle fuel market will continue to be driven by rising gasoline, as will diesel. Many estimates indicate that double the current lanthanum will be needed to meet the demand for hybrid vehicles that will use electric batteries to reduce gasoline consumption. But as lanthanum also plays an important role in telecommunications and the medical sector, demand for this metal is expected to remain very strong over the next few years.

Lanthanum is a rare earth metal, belonging to the category of lanthanides. It has many applications, such as in catalysts for oil refineries or in carbon lighting. Added in small quantities, it can decrease the hardness of hard metals such as molybdenum, ductility and malleability in steels. When lanthanum is added to the glass, it improves the alkali resistance. Lanthanum is also used in particular optical glasses such as infrared glasses and in lenses for cameras and telescopes, as well as in optical fibers. This rare earth is also a key component in laser lasers, nickel-hydride batteries, laptops, and nearly all portable electronic devices. It is also used in hydrogen cells in the automotive industry.

To date, almost 100% of lanthanum is mined in China, but this monopoly is very recent. In fact, in 2002 an environmental group, with the alleged financial support of China, managed to raise major protests in the United States and managed to obtain the closure of the two most important American producers, which until then had provided 54% of the needs of the United States. Since then, world supplies have been dependent on China which, as with other rare earths and rare metals, takes advantage of the monopoly to favor its own industries at the expense of Western ones, limiting exports to foreign countries.

Many objects that are part of our daily life contain lanthanum. For example, common lighters work thanks to lanthanum. In fact, the *mishmetal*, the pyrophoric alloy used in the stones of the lighters, contains from 25% to 45% of lanthanum.

The shortage of this important metal is a risk for the development of green technologies and its lack could also affect the availability of many technologies on which the West has bet for its future. Without lanthanum we will not have cars that do not pollute and the only slight consolation would be to see smokers with unlit cigarettes.

GREEN METALS TO REDUCE GREENHOUSE GASES

A small American start-up is bringing to the market a brand new technology that allows to produce metals without the emission of greenhouse gases and at much cheaper costs. It could be the beginning of a great change for various industrial sectors and, of course, for the environment.

Metal production is a major source of greenhouse gas emissions.

A small company, founded in 2008 and, since then, in silent work, is ready to bring to market its most innovative product, consisting of a device that looks like a brilliant ceramic tube, to produce many metals in a cleaner and cheap.

Infinum is a company that comes from Boston University (United States), whose work in recent years has focused on so-called rare earths, which include metals such as neodymium and dysprosium, metals used in the production of powerful magnets that can also work at high temperatures. But the new Infinium technology can be used to produce other metals, such as magnesium and aluminum.

The new Infinium process addresses a specific part of metal production: the transformation of minerals, partially processed and in the form of oxides, into metals.

The new technology can reduce processing costs by 30 to 50%.

A process that can traditionally be done by immersing the oxides in a bath of molten salts crossed by an electric current. Aside from the emissions associated with generating electricity, this process releases large amounts of greenhouse gases. In fact, one of the electrodes is generally made up of carbon which, reacting with oxygen, produces carbon dioxide.

The new ceramic material, in zirconium oxide, replaces the carbon electrode, completely eliminating polluting emissions.

Infinium has just started production using a machine that produces half a ton of rare earths per year and another machine capable of producing 10 tons per year will go into operation in September. The same process also works for aluminum, magnesium, titanium and silicon, metals for which the company plans to start production with the new technology by 2016.

Of course, the Infinium process is not a panacea to all environmental problems associated with metal production. In fact, it does not address pollution from extraction or separation of rare earth oxides from other materials contained in the mineral, but it is a step considered by experts to be very important, especially when it will be replicated on a large scale on the most important production plants of world.

Equally important is the economic aspect: according to Infinium, the new technology can reduce processing costs by 30 to 50%.

Making these metals much cheaper could, for example, make the electric car market take off in a short time, with obvious positive effects as regards global environmental aspects. Not to mention the greater convenience of using light metals to replace steel in cars, with a weight saving that would lead to a reduction in fuel consumption of at least 10 percent.

Finding an alternative to carbon has long been the "forbidden dream" of the entire metal industry and Infinium technology has all the characteristics to be able to make this dream come true.

RARE METALS

TUNGSTEN

Tungsten is essential for metalworking and therefore for the entire world industry but according to important investors it is also a great opportunity for the next few years.

The ore of tungsten was discovered in the 18th century in Sweden and has been present in many everyday objects since then, from metal cutting tools to light bulb filaments. When in 1781, Carl Wilhelm Scheele published the results obtained on the discovered mineral, it was called tungsten which in Swedish means heavy stone.

Tungsten is mined in various regions of the world, but China holds the top spot in world production with 75%. The other countries that produce it are Austria, Bolivia, Canada, Peru, Portugal, Russia, Thailand and many African countries.

Deposits of the mineral are present, for example, in areas where tectonic plates have collided to form mountains. The availability of tungsten is therefore very high, even if it is by no means easily accessible. The cost-effectiveness of tungsten extraction largely depends on the long-term trend of prices.

The applications of tungsten are very varied since hardness is one of its main characteristics, making it precious to be able to mold almost all materials, from metals to plastics and ceramics. About two thirds of the tungsten produced in the world is used in cemented carbide and in many applications in the construction and chemical fields. But we can also find it in devices that surround our daily life: devices for the vibrations of mobile phones, filaments of incandescent bulbs and solar panels.

There are some companies, listed on the stock exchange, which are engaged in this strategic sector and whose development prospects in the coming years seem very promising:

Woulfe Mining is the company that recently saw the famous tycoon Warren Buffett join its shareholders, with a 25% stake, engaged in an important project in South Korea.

EMC Metals owns the Springer mine in Nevada (United States), historically owned and operated by General Electric, which was forced to close due to too low prices.

Largo Resource Ltd is a Canadian company that produced 23,000 tons of tungsten concentrate in 2012 and plans to produce 42,000 tons in the next few years.

MOLYBDENUM

A metal, almost unknown to the general public, whose demand is expected to increase in line with the growth of the economies of emerging countries.

Try asking anyone what nickel is. He will probably tell you without hesitation that it is a shiny silver metal. But if you ask the same question about molybdenum you will not get answers, but simply stares into space.

Indeed, molybdenum is not a metal that occurs alone in nature. It is found in combination with other compounds. Its main use, as part of the steel production process, is actually hardly visible. This is not to say that molybdenum is not important. Quite the opposite: steel becomes much harder and highly resistant to heat and rust with the addition of a small amount of molybdenum.

If we take as an example a modern motor vehicle, whose weight varies from 3,000 to 5,500 kilograms, the molybdenum content does not exceed half a kilogram.

Molybdenum was discovered in 1778, but did not see significant industrial use until the end of the 20th century, when it was used in light bulb filaments. The demand for metal surged during the two world wars, when it was used to make the armor of tanks more resistant.

Molybdenum is often mined as a by-product of copper, but there are also mines that have very high concentrations of the metal. For example, Climax Molybdenum operates two molybdenum mines in Colorado (USA). The company claims that from about a ton of ore, it is able to extract about 2 to 3 kg of molybdenum.

Molybdenum (or moly as it is often called) is found in almost all types of steel, stainless steels, cast irons and super alloys, commonly used in aircraft, turbines and other high temperature and high stress environments. It has a high melting point (2.623 °C) and retains size and shape even when exposed to high temperatures. Thanks to these characteristics, there are very few substitutes for molybdenum as a binding agent for steel.

According to the International Molybdenum Association, about 80% of the molybdenum that is mined each year goes into steels. The remainder is used in chemicals, especially lubricants and paints. It is also used in oil refineries, specifically in catalysts to reduce the amount of sulfur in gasoline and diesel fuel.

In 2011, world production of molybdenum was around 250,000 tons, about 3% more than in 2010 and about 13% higher than in 2009. Demand remained strong despite a series of economic turmoil, including the crisis. eurozone debt and concerns about a Chinese slowdown, although oversupply could weigh on molybdenum prices in the coming years.

To date, molybdenum production is split between North America, China, and South America, which account for 33%, 31%, and 29% of annual production.

The demand for molybdenum is expected to grow in the coming years, in consideration of the urban development of all the BRIC countries (Brazil, Russia, India and China). In the long term, the adoption of more advanced technologies will generate a greater need for super alloys and high performance steels, for which molybdenum is indispensable.

Molybdenum also plays an important role in the energy sector, as it is used in significant quantities in oil and gas pipelines, as well as in high-performance steels for nuclear power plants.

THE LIQUID METAL

If watching the movie "Terminator" you thought that in reality a metallic robot would not be able to melt, you do not know gallium, a rare metal with surprising properties.

Gallium is a metal unknown to most people, as are its strange properties: it is a liquid metal (at room temperature) and was used in the first atomic bombs to stabilize plutonium.

Gallium is a scarce metal, but it is more abundant than better known metals such as antimony, molybdenum, tungsten and silver but, unlike these elements, gallium is not found in economic concentrations in natural minerals. The two main sources of commercial gallium are its extraction during the production of alumina (from bauxite) and from residues deriving from zinc oxides before electrolysis.

Much of the pure gallium production is in China, Germany and Kazakhstan. A significant percentage comes from secondary production, in particular from the recycling of gallium arsenide wafers (used to produce integrated circuits). The main centers of secondary production are Japan and North America. The global annual production of gallium is estimated at around 215 tons (2011), when only in 1986 the production was only 40 tons.

The market for this metal has evolved considerably in the last 2-3 years, with prices that have been affected by the global recession and the slowdown of the entire solar panel sector. The prices of gallium are now very close to the production costs. However, with demand driven primarily by light-emitting diodes (LEDs) used in iPads, cell phones and TV screens, many analysts think demand should start growing again in late 2012.

The prices of gallium doubled between the end of 2009 and mid-2011, but unlike precious metals which are highly correlated with the financial markets, rare metals follow the industrial demand and supply curve. With China refining about 70% of the world's gallium and quota quantities for exports, many investors are betting on gallium as the metal that will be driven up by demand growth, including that fueled by green technologies.

EVERYTHING, BUT REALLY EVERYTHING ABOUT MAGNESIUM

Materials science is inexorably making great strides and creating the best ground for great scientific breakthroughs.

Having to focus attention on a metal that could play the leading role among the materials that will have a substantial impact on the quality of our life in the future, there are not many doubts about which would be the choice: magnesium, the most available metal and lighter than known, 75% lighter than steel and 33% lighter than aluminum.

Even if magnesium is a little exotic and not too unknown metal, as for example are scandium or lithium or titanium, this does not mean that its properties are not extraordinarily useful for guaranteeing substantial technological developments in the not too long future. far.

Therefore, knowing this metal a little more is certainly not a bad idea.

According to historians, the first uses of magnesium date back to ancient Greece but production in commercial quantities only arrived in Germany in 1886. It is the eighth most abundant element on the earth's crust and is present in many minerals of which the most important from a commercial point of view are dolomite, magnesite, talc, carnalite, brucite and olivine.

THE MARKET

The production of magnesium compounds grew at a pace of just under 6% over the period 2002-2014, while demand grew at a slightly faster pace. A respectable growth rate but less strong than it could have been without the slowdown in emerging markets.

China controls the global magnesium market, with nearly 80% of all production. This situation is mainly due to the availability of abundant cheap labor, to permissive environmental standards and to very economical manufacturing processes.

The market for magnesium compounds is estimated at just over 7 million tons, according to 2014 data from the U.S. Geological Survey (USGS), with very broad uses including alloys, fertilizers, refractories, flame retardants and water purification.

But the most important market for this metal today is aluminum die casting alloys, which absorb two-thirds of all metallic magnesium.

According to a report by the United States Automotive Materials Partnership, an association between GM, Ford and Chrysler, by 2020 on every vehicle in circulation 110 kilograms of magnesium will replace 230 kilograms of steel and 40 kilograms of magnesium will replace 60 kilograms of aluminum, resulting in a overall weight reduction of 15%.

Of course, growth rates in the automotive sector are a key factor in increasing magnesium consumption in the future. If the demand for world cars were to grow in line with GDP growth, it would be reasonable to think that the demand for magnesium could follow this growth.

The second major use of magnesium is in refractories, directly related to steel production. In fact, the high melting point and its ability to remove sulfur from steel make magnesium indispensable for steel production. According to an estimate by Roskill, 50 grams of magnesium are used for every ton of steel produced. Again, much of the fate of the growth in steel demand is linked to the growth of China, one of the world's largest consumers.

PRICES

Speaking of prices, as with other industrial metals, there is no official market for magnesium. The price of the various magnesium compounds is determined on the basis of agreements between producers and consumers. In other words, prices are set with a "handshake" and this makes them particularly not very transparent. The most reliable indicative levels are the Chinese 99.8% magnesium export price and the European price ex Rotterdam.

Thanks to its diffusion on the earth's crust (magnesium is everywhere, even in seawater) and to its multiple uses, in metallic and compound form, there is a large number of companies, both public and private, participating in the market .

China holds a dominant position, with at least 50 Chinese companies involved in the magnesium business, 80% of which are located in the provinces of Shaanxi and Shanxi. China's largest magnesium producer is Shanxi Yinguang Magnesium Industry Group Co.

Outside of China, according to a Bloomberg estimate, there are 52 companies operating in the sector.

Looking towards the future, it is very likely that the structure of the market will undergo significant changes, since the Chinese government is paying increasing attention to environmental issues, with the consequence that many companies in the sector could be forced to close in the next years. A factor that will affect global magnesium supplies in the not too distant future.

A NEW SOURCE OF ENERGY: VANADIUM

Among the new sources of alternative energy, vanadium batteries stand out, large energy containers capable of satisfying the energy needs of entire buildings.

Not everyone knows that we could soon see a revolution in the energy sector, thanks to the new technology of vanadium flow batteries, energy containers for entire buildings and for electric vehicles.

The new technology is based on devices that use a completely new metal for these uses: vanadium.

Vanadium, a silver-colored transition metal, is currently used to galvanize steel, a process that makes it lighter and more resistant. But the advent of vanadium battery technology could rapidly change this landscape, leading to very rapid market growth.

This metal is not found in nature, but is produced from other minerals. China, Russia and South Africa are considered the leaders in world production. In 2013, China accounted for around 53% of all production, closely followed by South Africa (23%), while Russia produced around 10%.

The plant that produces the most in the world is the South African Bushveld Complex, which alone satisfies 23% of the global offer.

This metal is traditionally used to produce ferro-vanadium, a ferroalloy used as an additive to make steel lighter and resistant to corrosion and mechanical stress. The main products that contain it are components for airplanes, mechanical shafts, axles and various gears. With only 0.1% vanadium, the strength of steel doubles. For these reasons, the steel industry consumes 85% of all production, half of which ends up in the construction of various pipelines and pipelines.

Since the beginning of the 21st century, the demand for vanadium has grown steadily, with a slight decline of 15% in 2009.

In 2012, the world consumed about 80,000 tons of metal, and according to Roskill, demand will increase by about 28,000 tons between 2012 and 2017, largely due to an increase in steel production for the construction sector.

China, the largest producer of vanadium, is also the largest user. In the country, demand for this metal is expected to increase by more than 40%.

For all these reasons, observers are watching the advent of the new vanadium redox flow battery technology with interest and concern. Although work still remains to be done to make this technology commercially exploitable, the road now seems to have been drawn and, soon, a new actor will be fueling the world demand for vanadium.

And the concern of having reliable and stable sources is the main enemy of the new technology. Many believe that as soon as vanadium batteries are available, the problem of vanadium shortage will dramatically emerge, which will result in an increase in prices such as to make the use of these batteries uneconomical.

However you look at it, the future for vanadium producers looks full of positive surprises.

MATERIALS OF THE FUTURE: MOLYBDENITE

A recent discovery places molybdenite among the materials that could allow the development of some of the most futuristic technologies.

During the last century, the primacy for the material that most contributed to revolutionizing the way of life of mankind, according to many, is to be assigned to plastic. In the new century, what will be the candidate materials for the innovation Oscar?

Probably, at least in the current state of scientific discoveries, two materials are competing for primacy: graphene and molybdenite.

The former has been talked about a lot in some mass media, much less than molybdenite.

Molybdenite is the name given to molybdenum disulfide that is gaining increasing attention from researchers, thanks to its properties that can improve, for example, digital cameras and sodium ion batteries.

Put simply, molybdenite is a two-dimensional material with a thickness of up to a few nanometers. A bit like graphene with its extraordinary characteristics: a conductor of electricity better than copper, impermeable to gases and 200 times stronger than steel, but six times lighter.

Someone has called molybdenite a possible rival to graphene.

But, until now, no one has ever been able to measure the thermal conductivity of molybdenite. As experts explain, thermal conductivity is a crucial parameter for evaluating a material for use in electronics.

Angela Hight Walker and her team of researchers, from the Physical

Measurement Laboratory's Semiconductor and Dimensional Metrology Division (United States), have discovered how to measure the thermal conductivity of molybdenite using a technique called Raman spectroscopy. In doing so, they found that molybdenite is about a hundred times less efficient at conducting heat than graphene.

Some specific applications for the new material are already ready: new generations of electrodes for lithium-ion batteries and technologies for water splitting (i.e. the separation of oxygen and hydrogen), essential to power an economy based on hydrogen.

The future certainly has fascinating surprises in store for the use of this material, which we will still hear about.

TELLURIUM, A DUST ARRIVED FROM THE STARS

The discovery of tellurium on stars could reveal the origin of this rare metal on our planet. For geologists it would help to find new fields with which to cope with the growing world demand.

Scientists are trying to understand how tellurium could have formed on Earth, largely thanks to the discovery about a year ago of traces of this metal on three 12 billion-year-old stars.

Mistakenly, many people have come to believe that the rarest elements on our planet are rare earths, but in reality tellurium is much rarer.

It is scarcer than gold and any of the 17 elements that make up rare earths, having a density on the earth's crust of just 0.001 parts per million (ppm).

Furthermore, the world mining landscape sees a handful of countries with reserves of tellurium and among these countries only a few, three to be exact, extract and refine it to meet world demand, mainly due to its use in semiconductors, photovoltaic panels. and alloyed with other metals.

Sweden has many reserves and has contributed significantly to world production for many years. The Kankberg gold mine, managed by Boliden, has provided the largest amount of tellurium, increasing it over the past decade. According to the company's statements, 41 tons of tellurium were produced last year, representing just under 10% of global production.

Canada has estimated reserves of 800 tons and has recently expressed a willingness to explore other areas of the country in search of the mineral. Canadian reserves contain tellurium along with copper and gold.

China is the third, and last, country that supplies tellurium to the global market. Sichuan Apollo Solar S & T Co. (wholly owned by Apollo Solar), headquartered in Chengdu, Sichuan Province, is the leading mining company that extracts and refines this metal. It owns a mine in Dashuigou, in Sichuan province, and another in Majiagou. The latter two mines are the only ones in the world that exclusively produce tellurium as a primary product.

BECAUSE WE CAN'T DO WITHOUT COBALT

The numerous uses of this metal make it indispensable for our civilization and, at the moment, there are no valid alternatives to replace it.

Cobalt is a metal known since the time of the ancient Egyptians, who used it in 2000 BC. as a dye and, even today, it is among the most important minerals for man.

Its ability to maintain its resistance even at high temperatures makes it ideal for the production of cutting tools, super alloys, surface coatings, special steels and many other applications. Metal is also an essential element in the metabolism of humans and many animals; for example it is also used in fertilizing soils poor in cobalt to prevent the disease of weight loss in grazing animals.

One of the main applications of cobalt is in rechargeable batteries, the demand for which has increased dramatically over the past twenty years. If you think that in the mid-90s, only one percent of the cobalt extracted was used in electronics and compare it with today, where about 35% is used, you can better understand the growth trend in the use of this metal. The amount of cobalt used in electronics is set to rise over the next few years, due to the increase in demand for mobile phones, electric cars and all those devices that require rechargeable batteries.

The first versions of nickel batteries or lithium batteries had problems of short duration in the former and excessive reactivity in the latter (the batteries caught fire). The addition of cobalt has solved many of these problems and lithium batteries, for example, contain up to 60% cobalt per cell.

Finally a curiosity. There is an isotope of cobalt, called cobalt60, which is a powerful emitter of gamma rays and for this reason it is used in some nuclear weapons to increase radiation in the environment through the so-called fall-out, radioactive fallout in the form of ash and dust.

The increased demand for electronic devices based on rechargeable batteries has led to sharp increases in cobalt prices over the past 10 years. Despite efforts to find some substitute for cobalt to cope with rapidly growing demand, insiders predict that the demand for the mineral will increase dramatically over the next decade.

BECAUSE WE CANNOT DO WITHOUT AFNIO

Hafnium, a rare metal deriving from zirconium, could soon replace silicon in the production of microprocessors.

If you know what hafnium is, or you are a metallurgical expert or your memory of the periodic table of the elements is comparable to that of Giovanni Pico della Mirandola!

For most people, the name and uses of this element are totally unknown, although many objects that are part of our everyday life could not exist without the use of this rare metal.

Due to its unique properties, hafnium is used in many industrial processes. It is highly resistant to corrosion because it forms an oxide film on the surface, can be alloyed with other metals to create super alloys, is ductile, has a good ability to absorb neutrons, is resistant to acids and bases and has a point of very high melting and boiling. It is non-toxic in its elemental form and can ignite when powdered.

It is used in a variety of industrial and technological applications, including:

- Nuclear reactors
- Bulb filaments
- Jet engines
- Microprocessors

The world demand for hafnium is constantly increasing, not least because, in most cases, there is no substitute.

Hafnium is not found free in nature, but is found in zirconium deposits in percentages ranging from one to three percent. It is usually a byproduct of zirconium production. Both hafnium and zirconium are used in nuclear production, the former in control rods to prevent neutrons from escaping, the latter in fuel rods to conduct neutrons quickly and efficiently.

Hafnium is relatively abundant in the earth's crust, but due to production difficulties, the available quantity remains very limited, with high prices. The demand for hafnium is high due to its unique qualities and numerous industrial applications.

THE ALTERNATIVE NUCLEAR
ENERGY OF THE THORIUM

Thorium, a slightly radioactive metal found in rocks and soils, promises to be a more abundant, safer and cheaper substitute for uranium.

Worldwide, nuclear energy consumption is on the rise, while future uranium supplies remain highly uncertain.

However, there is a slightly radioactive metal, which could be a valid substitute for uranium: thorium.

Thorium is much more abundant on the earth's crust than its radioactive counterpart and is considered a safe and abundant alternative to uranium, with a very reasonable cost.

For this reason, some countries with large energy needs, such as China and India, have been interested in this form of alternative nuclear energy for decades now.

In 2013, the Norwegian company Thor Energy started producing power from thorium with an experimental nuclear reactor in Halden, Norway. The same company has created an international consortium, which also includes the nuclear giant Westinghouse and a division of Toshiba, to finance and manage research on the new energy source.

In addition to Thor Energy, other US, Australian and Czechoslovakian companies have engaged in research on thorium as a viable alternative to uranium. However, Thor Energy was the first to start producing nuclear power using thorium.

Unlike uranium, thorium is unable to develop a nuclear chain reaction, that is, in scientific terms, it is not fissile. However, if it is bombarded with neutrons from a fissile fuel, such as uranium-235 or plutonium-239, it converts to uranium-233, an excellent nuclear fuel. Once the process has started, it becomes autonomous with the fission of uranium-235 from thorium. As you can imagine, the details of the mechanism are much more articulated and complex than summarized in a few lines, but that's enough to get an idea of the difference between uranium and thorium in the production of nuclear energy.

The fact that thorium is not fissile by itself concerns a very important aspect in terms of safety: nuclear reactions can be stopped in an emergency.

But the characteristics that make the use of thorium instead of uranium so attractive are certainly its abundance (it also exists in Italy) and its cheapness. Thorium is present in small quantities in soils and rocks almost everywhere and it is estimated that it is 4 times more abundant than uranium. According to Reuters, the largest thorium reserves in the world are found in China, Australia, the United States, Turkey, India and Norway.

Furthermore, during a thorium-fueled nuclear reaction, most of the metal is consumed and, therefore, less nuclear waste is created. The only wastes that remain become non-hazardous after only 30 years, when the most dangerous of today's nuclear wastes must be kept safe for 10,000 years.

But there is more. Thorium could allow countries like Iran and North Korea to take advantage of nuclear energy without worrying the rest of the world about the secret development of nuclear weapons.

While succeeding in extracting energy from thorium in a cost-effective way still remains a challenge that will require major research and development efforts, this near-unknown metal could be the essential ingredient in combating the global threat of emissions from coal-fired power plants and the key to entering a new energy era.

LITHIUM FOR THE THIRD INDUSTRIAL REVOLUTION

The material that could revolutionize the entire modern transport system and portable electronics remains virtually unknown to the general public.

Will lithium be the engine of the third industrial revolution? The demand is inspired by the fervor surrounding the lithium market, especially for uses on electric vehicles.

But the market for electric vehicles and consumer electronics, in addition to being the most visible for the interest of the international press, is only the tip of the iceberg for the applications and sectors that are interested in lithium. For example, ceramics, lubricants and glass make up over 40% of the total lithium demand.

The electric vehicle market will still have to overcome several barriers before becoming the main driver for lithium demand. Many observers think it will take a few decades for the sector to realize its full potential. The costs of electric vehicles have taken the path of decline and in parallel new technological breakthroughs are emerging in the battery sector. Soon, energy storage facilities, such as storage depots and fueling stations, will become ubiquitous.

The new business models for battery manufacturers are in the direction of electricity storage and accumulation networks, in order to provide private users, industries and even governments around the world with the indispensable good for our society: energy. Many research laboratories are working hard to obtain energy efficiently and at low cost. Lithium is the key metal for many of these projects.

Recent statistical studies have shown that the lithium market should have no supply problems for at least the next 100 years.

The market is dominated by four major players, Rockwood Holdings, FMC, Sociedad Quimica y Minera and Talison Lithium, struggling with an overcrowded market and short-term oversupply. But the situation could undergo drastic changes in the next few years if global needs continue to grow. In Europe, for example, there are no important lithium deposits, excluding two small ones in Finland and Austria, while the largest concentrations are in Asia and South America.

Lithium is a rare metal used in industry and therefore is subject to fluctuations in global industrial production, but in the coming decades it will be an excellent opportunity for mining companies that will supply the raw material for the new industrial revolution.

INDIUM

Indium, a white and malleable metal, similar to aluminum or gallium, is a by-product of zinc. It is a rare metal, unknown to most people, although it is commonly used in all television sets and latest generation solar panels (CIGS). Historically, the drop in zinc prices has led to the closure of many mines, reducing the availability of indium. In addition, China, the main producer of this rare metal, has limited exports by further reducing the supply of indium.

According to Mathias Rueth, general manager of Tradium, Indian purchased for investment by individuals is on the rise: "We have customers who are shifting part of their money from precious metals to indium and gallium". During the Metal Bulletin's Ferroalloys conference, he said that it is estimated that about 5% of the 600 tons of indium produced annually ends up in the wallets of private investors.

80% of indium consumption is in LCD displays and flat screens for TVs. This sector is expected to continue to grow as the world population increases, while there is no pressure to seek replacement materials, both for quality and price reasons. In a television set, indium-based components account for less than 1% of the total price.

The new displays for Apple, produced by Sharp Electronics, use IGZO (zinc oxide, indium and gallium) technology that offers double the resolution and 90% energy savings compared to traditional displays.

World reserves are estimated at 11,000 tons, so in about 20 years there will be no more indium available. Will we go back to the years, when, until 1924, there was only one gram of pure indium in the world?

+ 335%, A BOOM CALLED NIOBIO

Is it possible that the prices of a company could grow by 335% in a few months, without maneuvers of financial speculation? This is what is happening to a small Canadian company that is preparing to start the first and only existing niobium mine in Europe and the United States.

When you are driving your car or taking a hot shower or sitting in the office, you should thank an unknown element of the periodic table: niobium.

Niobium, formerly known as columbium, is a rare metal that is added to steel to make it stronger and at the same time lighter and more flexible. It is increasingly used in the automotive sector, in the construction and energy sectors, in particular in gas and oil pipelines. It is also used in the nuclear industry, since it has a low cross section with thermal neutrons.

Niobium is an essential metal for Europe and the United States and has never been produced in these countries in the last 30 years. The whole world relies completely on Brazil for supplies and, for a small fraction, on Quebec (Canada).

China is the largest consumer of niobium in the world, due to the country's infrastructure boom. Even the most recent earthquakes, with dramatic damage in terms of loss of life, have highlighted the consequences of using poor quality materials in construction.

China needs to increase the use of niobium also to protect itself from the devastating effects of earthquakes but, unlike rare earths, graphite, zinc and iron ore, it cannot produce even a gram of niobium.

With this situation clear, the Chinese, Japanese and Koreans paid $ 2 billion to secure supplies from the Brazilian CBMM (Companhia Brasileira de Metalurgia e Mineração), the world's largest producer of niobium with 85% of all supplies. global. This shocked the West, which saw itself cut out of the negotiations.

In this context, a Canadian company based in Vancouver, NioCorp Developments Ltd, is developing the only primary niobium deposit in the United States, located in Elk Creek, Nebraska. Production of the metal is expected to start in a few months and could soon become one of the most important niobium production sites outside of Brazil.

These are the reasons that pushed the company's shares to grow 335% year to date, from $ 0.14 to $ 0.61. And, according to some analysts, the race does not end here and the next target is $ 0.80.

THE MINERAL OF FREE CLIMBING: MAGNESITE

For many athletes it is customary to have a bag containing a white powder with which to sprinkle the palms of the hands. But the importance of chalk is above all in the industrial field, where it is indispensable in many production processes.

What do free climbers, javelin throwers and weight lifters have in common?

All these athletes protect their most important tool, the hands, with a protective layer of chalk. That white powder with which the palms of the hands are covered before starting a climb or a race, introduced by the famous American climber John Gill in the 1950s to wipe sweat and to increase the grip of the fingertips and the hand.

But what are the properties of chalk that make it so important for many sports disciplines?

To understand this, we need to take a quick journey into the world of magnesium minerals, which will allow us to better understand the meaning and properties of many magnesite compounds that we have often heard about, without however knowing their meaning.

Although more than 80 minerals contain magnesium, only six of them are used for magnesium production. One of these materials, magnesite, is mainly used in modern production processes, ranging from paper to feed.

Also known as magnesium carbonate, this white crystalline mineral is found all over the world and its properties make it a valuable tool in a variety of industries.

In its purest forms, the mineral contains about 50% magnesium, which makes it ideal for the production of magnesium for use in aluminum alloys.

There are several methods for extracting magnesium from magnesite. However, the most common and least expensive method is to heat the ore between 1,200 and 1,600 °C, reducing the magnesium to a vapor. Once the vapor is cooled and condensed, pure magnesium is obtained. Unfortunately, this process is very harmful to the environment, as it contributes to global warming and produces a significant amount of waste materials.

But magnesite is also used to produce magnesium oxide, or magnesia, an important refractory material used in furnaces and kilns for many processes ranging from cement to non-ferrous metals.

Which are the main magnesite producing countries?

Once again, China dominates global magnesite production and, between 2011 and 2012, mined about six times more than Turkey, the world's second largest producer.

Thanks to this dominant position, China effectively controls the price of magnesium. Even if the recent regulations introduced in the country for greater respect for the environment should lead to an increase in prices, it will still be very difficult for all non-Chinese producers to be able to compete with China's low production costs.

All sportsmen, from those who do artistic gymnastics to those who jump with the pole or practice free climbing, will be happy to know that not only much of their clothing, but also the safety of their grip is 100% made in China!

COLTAN, A MINERAL THAT COSTS HUMAN LIVES

Conflicts and violence in East Africa have their roots in a semi-unknown mineral such as coltan, in great demand in the electronics market of developed countries, which until now have been willing to ignore its cost in human lives.

Someone may have already heard of coltan, a highly controversial mineral that comes from the Democratic Republic of Congo (DRC).

However, most people seem to be totally unaware of its existence, despite being one of the minerals most present in the daily life of all of us, from smartphones to computers, from medical equipment to electronic devices in cars.

Knowing more about this mineral can help to understand the bloody conflicts that afflict African populations, the cause of pain, suffering and poverty.

Coltan, the name of columbite-tantalite, is a mineral from which two important rare metals are extracted: niobium and tantalum. The first used 80% to manufacture low-alloy steels, while the second used throughout the electronics industry. Coltan is very rare, being present in the earth's crust in 2 parts per million.

About two-thirds of all tantalum produced in the world is used in the construction of electronic capacitors, key components of cell phones and electronic devices. It is tantalum that has contributed enormously to the miniaturization of modern electronic devices and, therefore, coltan is a key component of modern life.

It is mined, often by artisanal and improvised miners, in countries such as Brazil, Canada and Australia, although the main tantalum producer in the world turns out to be Rwanda which, in reality, acts as a figurehead for the Democratic Republic of the Congo, where the metal really comes from.

However, the hidden face of coltan is by no means uplifting. The ore is extracted manually, filtering sand and gravel until it settles to the bottom. Those who work in these mines are subjected to inhumane conditions, in shifts of no less than 12 hours, without any safety or health protection measures. A phenomenon that mainly concerns the mines in the Democratic Republic of Congo.

Furthermore, which is not new to our readers, coltan is one of the so-called conflict-minerals, that is minerals around which violent conflicts arise to control its extraction and trade, the proceeds of which in turn finance the purchase of weapons for armies of bandits and criminals. According to a recent estimate, the Rwandan army raised at least $ 250 million in 18 months through the sale of coltan. A problem that also affects tin, gold and tungsten.

Last but not least, the wild exploitation of coltan mines in the DRC has caused significant destruction of gorilla habitats. According to the United Nations Environment Program, the number of gorillas in eight national packs has decreased by 90% in the past 5 years and, to date, only 3,000 remain. It also appears that armed rebel groups and the miners themselves eat the meat of chimpanzees, gorillas and elephants in the Kahuzi Biega National Park and Okapi Wildlife Reserve.

For all these reasons, the European Parliament voted to introduce a ban on all those products that contain conflict-minerals, in the total disinterest of all public opinion, especially in Italy.

Do not be too surprised if when you turn on your smartphone you will feel something like a punch in the stomach ... it will mean that all the blood shed to give consumers in the most developed countries the latest miniaturized jewel of technology does not leave you completely indifferent.

THE MARKET

RARE EARTHS: JUNGLE OR MARKET?

Rare earths are not as rare elements as their name might suggest, but the market prices of these metals are something very complicated.

There are 17 rare earths and each of them is classified into different groups according to the type and form in which it occurs. Of course, the prices also change.

Argus Rare Earths, a rare earth information company, tracks 58 different prices, collected every two weeks. A daunting task for anyone who wants to figure out how to get around in what appears to be a wild jungle of prices.

However, with a little patience, it is possible to clarify by introducing some fundamental principles that govern the rare earth market and its prices.

First, it is important to know that the main driver of the market is China, the largest producer in the world, which produces over 90% of all rare earths. Thanks to this monopoly, in 2010 and 2011, when China reduced exports, the prices of rare earths shot up.

Consequently, the largest consumers of rare earths have begun to seek reliable supplies outside of China. This is anything but an easy undertaking, especially when prices have begun to drop significantly.

In 2014, the World Trade Organization (WTO) condemned Chinese restrictions on the export of rare earths and China decided to remove them starting in January of this year. Moreover, since May, the country has also eliminated export duties, condemning the prices of these metals to a further decline.

While some have hinted at the possibility that the Chinese monopoly may weaken in the near future, China is still the undisputed dominus of the market.

As for the prices, unlike gold or silver, it is not at all easy to find them as there is no official market, with the exception of the recently collapsed Fanya Metal Exchange. The only sources available are paid sources, such as that of Argus Rare Earths.

But rare earths, as we said earlier, are not all the same.

First of all they are divided between light and heavy rare earths and, in principle, the latter are the most requested. Not for this reason, among the light rare earths, it can be said that there are metals of little importance as in the case of neodymium and praseodymium, indispensable for the manufacture of magnets together with dysprosium. Dysprosium is very expensive since, in metallic form, it is currently worth $ 270 per kilogram, while dysprosium oxide is worth $ 210 per kilogram.

Instead, the price of metallic cerium is around $ 5.40 per kilogram. Cerium is the most abundant of the rare earths, even more abundant than copper.

Both cerium and lanthanum, both used in steel production, are currently in oversupply. As well as yttrium which is quite cheap ($ 4.40 per kilogram), as opposed to darn rare and expensive europium and terbium. Metallic terbium is worth $ 520 per kilogram and terbium oxide $ 380. Metallic europium $ 345 and europium oxide $ 110 per kilogram.

Such a variety entails the need to physically separate the 17 rare earth elements from each other, an operation that is not at all easy, also complicated by the presence of a number of other impurities such as uranium and thorium, which are difficult to dispose of.

Although it is customary to use an average of rare earth prices to get an idea of the general trend, it is impossible to understand the trend except by separating the different metals, which often follow a logic of supply and demand. very different from each other.

As many mass media did during the years of sensational price hikes, not understanding the differences between one element of rare earths and the other, there is the risk of making gigantic blunders in formulating forecasts on the evolution of land prices. rare as if they were a single metal.

THAT'S WHY COBALT PRICES WILL GO UP

Today, in the tumultuous commodity markets, it is very difficult for an investor to choose a metal to bet on.

Precious metals are doing very badly and the forecasts are not too encouraging for the foreseeable future. Not very different is the scenario of industrial metals, some of which have been dramatically affected by China's economic slowdown.

In the midst of this confusion, some believe that some good opportunities can be found among the so-called rare metals, or critical metals. This is the case of 3 metals, which are part of the supply chain of those who produce batteries: lithium, graphite and cobalt.

We will focus our attention on cobalt, with a brief overview to understand why some analysts are optimistic about this metal.

It is easy for anyone to imagine the positive impact that electric vehicles will have on cobalt. The construction of Tesla Motors's new lithium battery Gigafactory, with an investment of $ 5 billion, has attracted international attention not only to lithium and graphite, but also to cobalt, indispensable in the production of batteries.

Although the quantities of metals needed by Tesla's new factory are not yet known, some analysts have formulated estimates that 7,000 tons of cobalt will be needed per year. Furthermore, Tesla is not the only company to have plans to build battery factories for electric vehicles.

Unquestionably, demand seems destined for growth in the coming years, much higher than the current one, which is 6% per year. But what is happening on the supply side?

Up to now, cobalt supplies have never been a problem, but things seem to be at the beginning of a change. Glencore recently suspended production for 18 months at Katanga Mining and Mopani Copper Mines, two copper mines that produce cobalt as a by-product in an amount of approximately 16,000 tons per year. Considering that world production in 2014 was 112,000 tons, you can understand the impact that the disappearance of Glencore's cobalt will have on the market.

Still on the supply side, another endemic problem afflicts the market. The most important cobalt producing country in the world is the Democratic Republic of the Congo, known for its political instability. Even if today there are no particular problems in this regard, the threat of sudden supply interruptions is a sword of Damocles that hangs over the entire market.

For all these reasons, a growing number of analysts foresee a bullish future for cobalt. When this happens, however, it remains among the famous "100 million dollars" questions!

THE SCANDAL PRICES OF SCANDIUM

Following the strong increases of the last period, many investors would like to know the prices of scandium but, as is the case for other critical metals, the prices are not at all easy to find.

Lately, scandium has been attracting the attention of many investors.

The potential demand for this rare metal is enormous, mainly due to its use in some aluminum alloys and for fuel cell applications, the batteries that represent the future of transport.

According to the US Geological Survey, supplies and consumption of scandium around the world range from 10 to 15 tons per year, but consumers of this very scarce metal would be willing to buy more if some new source of supply were available.

As you can imagine, this market is not easy to approach even for the most experienced traders and, like most critical metals, prices are very difficult to find.

Copper, silver and gold, just to give examples, are listed on regulated markets and the prices are in the public domain. For scandium things are different. There are no forward or futures contracts and therefore there is no market where buyers and sellers can set a price. Prices are established between individual buyers and sellers during their private negotiations.

Furthermore, there is no standard qualitative reference for the metal and, therefore, the prices refer to different purities which can reach 99.9% for scandium oxide used in electronic applications, but which are much lower for the production of aluminum alloys.

However, there is not only scandium oxide, as the market also requires scandium chloride, fluoride and acetate.

But let's get to the part that interests investors most: the price.

According to the latest available data from the US Geological Survey, which publishes price estimates for this metal almost every year, scandium oxide of 99.99 purity was worth $ 5,000 per kilogram. But scandium oxide 99.9995 was even worth $ 6,000 per kilogram.

Considering that prices have risen considerably since these latest surveys, it is natural to ask whether these levels are too high.

However, for scandium buyers, the main problem is not prices, but finding a constant and reliable supply of metal.

Therefore, even if some analysts indicate that prices could slow down in the coming years, it is not too sure that this will happen anytime soon.

BILLIONARY BUSINESS IN THE RECYCLING OF RARE METALS

An unexplored open pit mine of rare and precious metals. Here's what the rare metal recycling business could become in a few years.

Hundreds of millions of pounds in rare and precious metals are thrown away every year in England, where cell phones and old computers are scrapped. The UK government is rolling out a new plan to help UK companies benefit from the multi-billion dollar market for reuse of these metals.

Faced with a growing global demand for consumer goods and a foreign dependence on these metals, the secretary for the environment Caroline Spelman, has recently published a plan, the Resource Security Action Plan, to ensure that companies Brits may be less vulnerable to price and supply changes.

China produces over 95% of rare earths, while Russia and Congo are leaders in the production of other rare metals essential for the technologies we use every day, such as mobile phones. 80% of the general managers of companies, recently interviewed by the Association of British Industrialists, stated that the shortage of rare metals represented a big risk for their company's business in 2012.

Between now and 2020, England will have 12 million tons of scrapped electronic equipment, containing rare and precious metals in considerable percentages: 63 tons of palladium (worth 1 billion pounds) and 17 tons of iridium (worth 380 million pounds).

The Resource Security Action Plan provides funding of £ 200 million to grow companies that find new reuse and recycling methods for these metals. Support and investments are planned to encourage, help and support projects to start new businesses that provide the recycling of rare metals.

Metals used in major tech consumer goods include:

- cell phones: gold, antimony, palladium, beryllium, gallium and platinum (cobalt in batteries);
- laptop: cobalt and nickel (hard disk) and neodymium;
- mp3 player, headphones and speakers: neodymium;
- rechargeable batteries: cobalt;
- hybrid vehicles: lithium and rare earths;
- televisions, computers and other electronic devices: indium, cerium, lanthanum, praseodymium;
- electronic boards: cobalt, gallium, lithium and platinum;
- jewelry: platinum;
- vehicles: platinum, palladium and rhodium;
- medical devices: platinum.

But there is also a multi-billion dollar opportunity in the massive amount of precious metals being lost due to the way we approach products today that people no longer want.

THE FUTURE BODES WELL FOR URANIUM

The uranium market is used to thinking over long periods of time and observers expect current prices to at least double over the next ten years. Here because...

The recent past of uranium has been characterized by an excess of stocks which, added to new productions, has led to an offer exceeding the market demand.

To better understand the functioning of a market such as that of uranium, it is essential to investigate the mechanisms that govern it. In fact, there are two main sources of uranium: mines, which contribute about three quarters of global demand, and secondary sources. Secondary sources are government stocks and re-enriched sterile uranium.

Currently, both the US and Russia are selling uranium on the market from their own stocks.

As for prices, the spot price of uranium is around $ 37 per pound (U3O8), while the price for long-term contracts reaches $ 49 per pound. Low prices, which certainly do not encourage production or the start-up of new plants.

The current annual mining production amounts to 68,000 tons, a very high quantity compared to the last decade and which will force the market in the coming years to produce no more than 11,000 tons per year, if prices do not change.

Looking instead at the demand for nuclear power, France and the United States occupy the first places, followed by Russia, China and South Korea.

However, China has plans to start several nuclear reactors, as well as Russia, India and South Korea, which will increase the demand for uranium as a nuclear fuel over the next five years. For this reason, analysts believe that the next 15 years will see a market growth rate of 3%, the largest growth ever recorded since 1970 and leading to a significant deficit by the end of the decade.

What about the prices? Expectations are that uranium prices will hit $ 70 per pound over the next two or three years.

Furthermore, an event that has never occurred before will take place during this year. The Uranium Participation Corp (UPC) investment fund, a fund dedicated to physical uranium, will make its first major purchases on the market. The investment is likely to be worth around $ 200 million, equivalent to 900 tons of uranium.

However, the short-term market focus is on the imminent restart of Japanese nuclear reactors which, as expected for months, could ignite the market and lead to a significant rise in uranium prices.

www.ingramcontent.com/pod-product-compliance
Lightning Source LLC
Chambersburg PA
CBHW070317240526
45467CB00046B/1419